In the same series:

EINSTEIN & RELATIVITY
NEWTON & GRAVITY
CRICK, WATSON & DNA
PYTHAGORAS & HIS THEOREM
HAWKING & BLACK HOLES
OPPENHEIMER & THE BOMB
GALILEO & THE SOLAR SYSTEM
ARCHIMEDES & THE FULCRUM

Turing & the Computer

PAUL STRATHERN

Published in the United Kingdom in 1997 by
Arrow Books

5 7 9 10 8 6 4

First published in the United Kingdom
in 1997 by Arrow Books

Arrow Books Limited
Random House UK Ltd
20 Vauxhall Bridge Road, London SW1V 2SA

Random House Australia (Pty) Limited
20 Alfred Street, Milsons Point, Sydney,
New South Wales 2061, Australia

Random House New Zealand Limited
18 Poland Road, Glenfield
Auckland 10, New Zealand

Random House South Africa (Pty) Limited
Endulini, 5a Jubilee Road, Parktown 2193, South Africa

Random House UK Limited Reg. No. 954009

A CIP catalogue record for this book
is available from the British Library

Papers used by Random House UK Limited are natural, recyclable products made from wood grown in sustainable forests. The manufacturing processes conform to the environmental regulations of the country of origin

Typeset in Bembo by SX Composing DTP, Rayleigh, Essex
Printed and bound in the United Kingdom by
Cox & Wyman Ltd, Reading, Berkshire

ISBN 0 09 923782 2

CONTENTS

INTRODUCTION

The development of the computer could well prove one of humanity's greatest technological achievements. The computer may yet rank alongside the use of fire, the discovery of the wheel and the harnessing of electricity. These previous advances harnessed basic forces: the computer harnesses intelligence itself.

Over 90% of the scientists who have ever lived are alive now, and the speed of their work is multiplied daily by computer. (The mapping of the human gene will probably be completed *half a century* earlier than was predicted on the discovery of its structure, all because of computers.)

But hopes should not be raised too high. Similar expectations accompanied the development of the steam engine less than 150 years ago. And the slide rule lasted less than a century. The advance which will render the computer redundant is only inconceivable because it has not yet been conceived.

Even before the first computer had been developed, we knew its theoretical limits. We knew *what* it could compute. And even as the first computers were being put together, the potential *quality* of their power was understood: they could develop their own artificial intelligence. One man was responsible for both these ideas – his name was Alan Turing.

An idiosyncratic fellow, who came to regard himself as something of a computer, Turing also worked on the Colossus calculating machine, which cracked the German Enigma codes during the Second World War. Like Archimedes, Turing was forced to set aside a brilliant mathematical career in order to try and save his country. Archimedes failed, and was put to the sword by a Roman soldier. Turing succeeded, and his grateful country prosecuted him for homosexuality.

Turing was largely forgotten after his untimely death, but is now increasingly recognised as possibly the major figure in the history of the computer.

THE BC ERA: COMPUTERS BEFORE THEIR TIME

The first computer was of course the abacus. This method of calculation was invented even before the wheel. (Our desire not to be cheated is evidently deeper than our wish to travel in comfort.) Archaeological evidence indicates that a form of abacus was being used around 4000BC in both China and the Near East. It seems to have evolved independently in these two regions. Some have suggested that this indicates the primacy of mathematics: the urge to calculate as a seemingly inevitable function of the human condition.

Abacus derives from the Babylonian word *abaq*, which meant 'dust'. Scholars have come up with characteristically ingenious explanations for this glaring *non-sequitur.* According to one version all calculations were originally done in the dust, so dust became the name for any form of calculation. Or alternatively, the method of calculating used by the abacus was first drawn in lines and scratches in the dust.

In fact, the abacus is not strictly speaking a computer at all. The actual calculating is done by the operator of the abacus, who must have in his head the programme (the necessary mathematical steps).

Computer or no, the abacus and its human programme were certainly used for computing well into the Middle Ages, throughout Europe and Asia. Then came the introduction of the zero into mathematics, which threw a spanner in the works where the abacus was concerned. As a result, serious mathematicians quickly disdained this infantile aid. Yet for centuries afterwards the abacus continued to be used as a calculator, cash register, computer and whathaveyou (or whathaveyounot). Indeed, to this day the abacus still plays a pivotal role in the local economy of parts of central Asia and Russia.

The earliest known calculating machine remains a complete mystery. In 1900 Greek sponge divers discovered off the tiny Aegean island of

Antikythera an ancient shipwreck dating from the first century BC. Amongst the broken statuary and jars were found some pieces of corroded bronze, which appeared to be part of a machine. Not until 50 years later did scholars manage to work out how these pieces fitted together, and construct a working model. This turned out to be a type of astronomical calculating machine, which functioned just like a modern analog computer, using mechanical parts to make calculations. A crank could be turned, which operated gears; this in turn operated dials, from which it was possible to read the position of the sun and the planets in the zodiac.

What makes this find so astonishing is that it is unique. Nothing remotely like it has ever been discovered from this period. There is no mention of such a machine or anything similar anywhere in ancient Greek literature. No philosopher, no poet, no mathematician, no scientist or astronomer makes any reference to such an object. And according to our present knowledge of ancient Greek science, there was no tradition or expertise capable of producing such a machine. The first computer appears to have been a freak construction, perhaps as a toy, by some unknown mechanical genius who simply vanished from history. A freak without influence, it disappears like a comet. Then nothing – for over one and a half millennia.

The first 'real' mechanical calculating machine is generally reckoned to have been that produced in 1623 by William Schickard, professor of Hebrew at Tübingen University. Schickard was a friend of the astronomer Johannes Kepler, who discovered the laws of planetary motion. Kepler awakened a latent interest in mathematics in the Hebrew professor, whose ability to calculate had evidently grown a little rusty over the years. So he decided to construct a machine to help him with his sums. Schickard's machine was described as a 'calculating clock'. It was intended as an aid to astronomers, allowing them to calculate the ephemeris (the future positions of the sun, moon and the planets).

Unfortunately we will never know whether this machine worked, or precisely how it was intended to work. The first and only prototype remained incomplete when it and Schickard's blueprints were destroyed by fire during the

Thirty Years War. Schickard was thus reduced to a mere historical footnote, rather than the inventor of the most important technological advance since the harness.

We *do* know that Schickard's machine was a forerunner of the digital computer, whose input is in the form of numbers. For the other type of computer, the analog computer, the input (and output) numbers are replaced by a measurable quantity – such as voltage, weight or length. The last of these was used in the first analog computer, the slide rule, which was invented in the 1630s. The simplest slide rule involves two rulers, both marked with logarithmic scales. By sliding the rulers so as to place one number against another, it is possible simply to read off multiples and divisions.

The slide rule was invented by William Oughtred, whose father worked as a scrivener at Eton, teaching the illiterate pupils how to write. His son took holy orders as a priest, but followed in his father's footsteps by doing some tutoring on the side. In the 1630s he produced the first rectilinear slide rule (ie, with two straight rulers). A few years later he came up with the idea of the circular slide rule (which has a moveable circle within a ring, instead of sliding rulers). Unfortunately one of his pupils purloined this idea and published it first, claiming the discovery as his own. Oughtred was not pleased; but he was to end his days a happy man. A dedicated Royalist, he is said to have died in a 'transport of joy' after hearing that Charles II had been restored to the throne.

The primitive early slide rule was developed over the years into a device capable of sophisticated calculation. Amongst those who contributed to this development was James Watt, who adapted it for use in calculating the design of his pioneer steam engines in the 1780s. A further advance was made by Amadeé Mannheim, a French artillery officer. He designed an advanced form of slide rule whilst still a student, thus enabling him to achieve the outstanding exam results which launched him on a brilliant career in military education. It was Mannheim's version of the slide rule which achieved such widespread use during the first half of the 20th century – the *de rigeur* sartorial accessory in the top pocket of every white-coated scientist.

But back to the digital computer. The next advance in this field was made by the 17th century French mathematician Blaise Pascal, who was coincidentally born in 1623, the same year that Schickard had invented the original 'calculating clock'. Pascal's father was a royal tax official – who found it difficult enough to collect the cash, let alone produce the necessary accounts for the royal exchequer. In order to assist him, his precocious young son set about designing an accounting machine. By the age of 19 he had produced a working model. The numbers were entered into the machine by dial wheels connected to rods with toothed cogs and gears. Pascal's machine was capable of addition and subtraction, involving figures up to eight digits. This machine was extremely complicated, stretching current mechanical techniques to their limits – and sometimes beyond. The machine was beset with teething problems. But Pascal was a perfectionist and claimed to have made 'more than 50 models, all different'. Pascal was not only a great mathematician, he was also the finest religious philosopher of his age. Plagued by failing health, his religious zeal increased in inverse proportion to his health. But he remained a mathematician to the end, even reducing faith to mathematical probability. In his view, although one could calculate the odds against God's existence, it was better to gamble that He did exist – because you had nothing to lose if He didn't exist.

Seven of Pascal's machines still survive: ingenious masterpieces, which incorporate several principles still used in mechanical computers. A number of the surviving Pascal machines are still in working order – though no one has yet discovered how to use them to calculate the odds against God's existence.

The next significant advance in the digital computer was achieved by the German philosopher Gottfried Leibniz, who was the Leonardo of his age. Amongst many other things Leibniz was to produce no less than two philosophies (one optimistic, the other pessimistic), a detailed plan for the invasion of Egypt, a 15-volume history of the House of Hanover – and a calculating machine which far exceeded Pascal's.

Leibniz' interest in calculating machines was more than just practical. Whilst still at university he produced a paper explaining the theoretical basis of any calculator, and what it could do. (A work which pointed the way ahead to Turing's seminal ideas on this subject almost 300 years later.) Around the same time he also invented a binary mathematics, such as the one that was to become the language of digital computers – though he didn't combine the two.

Leibniz produced his calculating machine in 1673, after having seen one of Pascal's machines in Paris. Unfortunately Leibniz was broke at the time, and his efforts were hamstrung by the need to make his machine commercially viable. (Pascal's was far too complex to be produced by anyone but himself.) As soon as Leibniz had finished his machine, he set off across the Channel to demonstrate it to the Royal Society in London. Its members proved uninterested, and he dropped the project at the prototype stage.

Despite these limitations, Leibniz' machine was remarkable. Like Pascal's it was driven by a succession of toothed cogs. But it was capable of much more than Pascal's. Right from the start it could multiply (by repeated additions), but he soon incorporated devices which enabled it to divide and also to calculate square roots.

Leibniz saw a great future for calculating machines – though he never again found time for practical endeavours in this field. This didn't stop his ever-active mind from thinking about calculating machines, and the role they might play in a future world. In his view, all ethical disputes would one day be solved by calculating machines. One would simply feed in the different arguments, and the machine would 'calculate' which argument was superior. (Though the precise basis for these calculations remained in the same category as calculating the odds against God's existence – a mystery to all but the genius who conceived of it.)

In a similar vein, Leibniz also predicted that calculating machines would soon render judges redundant: the law courts of the future would be presided over by calculating machines – which would produce both the verdict and the appropriate sentence. Such astonishing prescience may conjure up the ultimate computer horror story, but Leibniz saw it very differently. He was essentially an optimistic soul, believing that 'all is for the best, in this the best of all possible worlds'. Had he devoted more of his exceptional energies to producing calculating machines, there's no telling what kind of possible world they might have produced.

The next significant advance in the field was made by a complete outsider. Joseph Marie Jacquard was a French technician in the weaving business. In the early years of the 19th century he put together an innovative loom where the weaving pattern was controlled by punched cards. Thus began the idea of programming a machine, though Jacquard had little idea of the significance of his invention. He refined this idea, but it worked too well. His machines provoked riots in Lyons during the 1820s, when redundant loom workers stormed the factories and many of his machines were destroyed. Jacquard's method is still used for weaving complex patterns.

Intricate mechanical calculating machines, the idea of programming, a theory of computable numbers – the basic elements of the modern computer were beginning to appear. But it took a genius to recognise how these disparate elements could be combined. Charles Babbage is generally regarded as the father of the computer.

Like many a genius in the practical field, he was hopelessly impractical in any real sense of the word. But his discoveries and achievements were a century ahead of their time.

Babbage was born in 1791, and inherited a considerable personal fortune. An affable young man, he quickly showed exceptional promise at mathematics. He campaigned successfully for the introduction of Leibniz' calculus notation in Britain. (British mathematicians had patriotically insisted upon using Newton's original but inferior notation, thus largely isolating themselves from a century of Continental advances.)

Babbage then turned his attention to another bugbear which was hampering British scientists – notably, the multiple errors which littered all printings of astronomical and mathematical tables. For example, the first edition of the *Nautical Ephemeris for Finding Latitude and Longitude at Sea* was found to contain over a thousand errors!

Babbage decided there was only one answer to the problem of faulty tables. It was necessary to construct a large, all-purpose, infallible calculating machine. After successfully applying for a government grant, Babbage set about constructing his celebrated 'Difference Engine No. 1'. This was a colossally ambitious undertaking. Babbage's machine was to be capable of calculating to 20 digits; it would also be able to store a series of numbers and undertake additions of these numbers. The machine's calculations could be limited to additions, because it would employ the method of multiple differences. This makes use of polynomials (algebraic formulae consisting of several terms), and the fact that those have a constant difference. Put in its simplest form:

Where $f(x) = 2x + 1$

x =	1	2	3	4 . . .
$f(x)$ =	3	5	7	9 . . .
differences =	2	2	2	2 . . .

Needless to say, it is not so easy with more complex functions. But here a constant difference may be found in the differences between the differences (or the differences between the differences between the differences). In most cases, if a polynomial has a term x^n, then n differences must be calculated before a constant difference is found. To work out a polynomial for a succession of values of x, as is required when computing tables, it is much easier for a machine to add the constant difference and then work back adding differences – rather than become involved in a complex series of multiplications. And functions like logarithms and trigonometric functions, which don't work like this, could be reduced to closely approximate polynomials.

Like its predecessors, the 'Difference Engine No. 1' used toothed wheels and operated on the decimal system – but its construction vastly exceeded all others in complexity, requiring a succession of innovations in mechanical engineering.

But Babbage was well up to this task, being a past master at extemporising on the hoof. As his machine grew, so he kept having brilliant ideas for innovative features, incorporating them as he went along. The 'Difference Engine No. 1' was begun in 1823, but was never to be finished. After ten years work, Babbage had enlarged his original plans into a machine of 25,000 parts (only 12,000 of which had been made), and the cost had spiralled to £17,470 (in those days, enough to build a couple of men-o'-war). Babbage contributed large sums out of his own pocket, but the government decided to call a halt to the scheme. It was better to invest in a navy, rather than in a machine which might end up contributing to the national debt in figures which it alone could calculate. Despite all these difficulties, by 1827 Babbage had used the only working part of his machine (which consisted of a mere 2000 parts) to calculate log tables from 1 to 108,000. This working part of the 'Difference Engine No. 1' is generally regarded as the first automatic calculator. The figures were fed in, and the answers came out in printed form (thus eliminating further the possibility of human error).

But this was only the beginning, where Babbage was concerned. By the 1830s he already had plans for a 'Difference Engine No. 2'. This concept was a significant advance in computing techniques. It was to be the first analytic machine: one whose function was controlled by an external programme. Babbage had heard about Jacquard's idea of punched cards to control the mechanism of a machine, and decided to incorporate this idea into his own machine. This would enable it to perform any particular arithmetical task according to inserted instructions from cards with holes punched in them. Like the first Difference Engine it would also have a memory where it could store numbers, but the new machine would be able to carry out a sequence of operations involving these stored numbers. Babbage had come up with the essential features of the modern computer.

The central mill, to which these features would be attached, was to be the *pièce de resistance*. This was to contain a thousand axle rods with no less than 50,000 geared wheels, and would be able to calculate with 50 digit numbers using the decimal system.

Unfortunately the government refused to be cowed by these awesome possibilities, and decided against a second attempt to bankrupt the exchequer. By now the strain of many years of hard work without result had played ravages with Babbage's character. The personable young man of Cambridge had become an irascible old fart stalking the streets of London. He got a bee in his bonnet about the noise created by street musicians which 'not infrequently gives rise to a dance by little ragged urchins, and sometimes by half-intoxicated men, who occasionally accompany the noise with their own discordant voices . . . Another class who are great supporters of street music consists of ladies of elastic virtue and cosmopolitan tendencies, to whom it affords a decent excuse for displaying their fascinations at their own open windows.' Babbage mounted a campaign to have all street musicians banned, claiming that they prevented him from working in peace. The street musicians retaliated by gathering outside his window. Babbage recorded that 'upon one occasion a brass band played, with but a few short intermissions, for five hours'.

Babbage had by now spent much of his personal fortune following his dream of difference engines. For several years he was assisted in his endeavours by Ada, Lady Lovelace, the daughter of the poet Byron and one of the finest women mathematicians of her time. (Her role in computer history received its supreme acknowledgement when the US Defense Department named its programming language ADA after her.) Lady Lovelace also assisted Babbage in an optimistic attempt to recoup his fortunes. Together they devoted much time and energy trying to work out an infallible gambling system for horse-racing. Alas, during the practical trials this system proved almost as costly as a difference engine.

Despite such setbacks Babbage also found time to invent the locomotive cowcatcher and discover how tree rings could be read as climatic records. After Babbage died in 1871 his working designs for the 'Difference Engine No. 2' lay forgotten for many years. Subsequently the mill of the world's first analytic engine was constructed according to modified plans for the 'Difference Engine No. 2'. This magnificent three ton construction can now be seen in all its glory in the Science Museum, London. And it works. (At its trials it was set to calculate 25 multiples of π to 29 decimal places – a task which its 50,000 geared cogs digested with consummate ease.)

Babbage had set out the basic features of the modern computer, but his machines suffered from one crucial drawback. They operated in decimal mathematics. This was to be remedied as a result of work done by George Boole, one of his contemporaries. Boole was born in 1813, the son of a Lincoln cobbler. Though almost entirely self-taught, he demonstrated such intellectual acuity that he was appointed Professor of Mathematics at Queen's College, Cork – where he eventually married Mary Everest, niece of the man after whom the mountain is named.

In 1854 Boole published his 'Investigation of the Laws of Thought', which introduced what is now known as Boolean algebra. In this work, Boole suggested that logic is a form of mathematics, rather than philosophy. Like geometry, it is based on a foundation of simple axioms. And where arithmetic has primary functions such as addition, multiplication and division, logic can be reduced to operators such as 'and', 'or', 'not'. These operators can be set to work in a binary system. (Where the decimal system has ten digits, binary works in the same way with just two.) The 'true' and 'false' of logic are reduced to the 0 and 1 of binary. Binary algebra thus reduces any logical proposition, no matter how many terms it may contain, to a simple sequence of binary symbolism. This could be contained on a strip of paper where the binary algebra is reduced to a sequence of holes (and no holes). In this way, an entire logical 'argument' or programme could easily be fed into a machine.

With binary digits, machines could follow logical instructions and their mathematics was perfectly adapted to the on/off of electrical circuitry. As a result, the binary digit (or bit) was eventually to become the fundamental unit of information in computer systems.

But for the time being the separate advances of Babbage and Boole lay unrecognised. As far as the world was concerned, the next significant advance came from Herman Hollerith, a US statistician. Hollerith developed a 'census machine' that could read cards with up to 288 holes, allowing it to store this information. His electromechanical machine was able to read as many as 80 cards a minute. When it was used for the 1890 US census, Hollerith's machine managed to process all the data in six weeks. (The previous 1880 census had taken three years.) In 1896 Hollerith went into business, setting up his Tabulating Machine Company, which later evolved into the International Business Machine Corporation (IBM).

The elements necessary for the modern computer (including commercial exploitation) had now been discovered. All that remained was for someone to work out what it could do – its theoretical possibilities and limitations. This was to be done by Alan Turing.

LIFE & WORK OF AN ENIGMA

Alan Turing was born in London in 1912, into an upper middle-class English family. His father was in the Indian Civil Service, and his mother was the daughter of the chief engineer of the Madras Railway. In 1913 his mother and father returned to India – leaving the infant Alan and his five-year-old brother to be looked after by a retired colonel and his wife at St Leonards-on-Sea in Sussex.

In those days respectable English parents thought nothing of abandoning their children in this fashion. Even those who couldn't vanish to the colonies employed nannies and boarding schools (from the age of seven to eighteen) to make sure their children were seldom seen and never heard. This early abandonment had as little effect on Alan's brother John as it did on most of his middle-class generation – who all turned out to be typical English public schoolboys. (Only in our more demanding era has this distinctive species come to be regarded as emotionally crippled.) But Alan Turing turned out to be a normal child – and as such was profoundly damaged by this early experience. He developed a pronounced stutter, was to become self-sufficient to the point of eccentricity, and found himself incapable of joining in the charade of social mores.

When his mother returned for a longish visit in 1916, young Alan reacted with mixed feelings – which were to remain with him throughout his life. He loved his mother dearly, but also found her impossible. Mrs Turing for her part seems to have been impossible but unlovable. Her main concern in life was that Alan should appear respectable. The imp inside him was to make this too impossible.

At boarding school Alan began as he meant to go on. He was the one who was always scruffy, had ink on his fingers, and didn't fit in with the others. Worse still, he didn't seem to want to fit in. He was shy, lonely, and his stammer only made matters worse. Young Turing minor showed no promise whatsoever. He had difficulty learning to read and write. Then one day he decided he wanted to read, and simply taught himself in three weeks. By the age of 11 he had developed a passion for organic chemistry, but remained blithely uninterested in other subjects and couldn't even do long division.

When Mr and Mrs Turing returned from India for good, they decided to settle at Dinard, in Brittany, in order to avoid paying tax. This was an almost unselfish act on their behalf. You simply weren't middle-class if you didn't send your sons to public school – which cost more money than the Turings could easily afford. This social stigma had to be avoided at all costs, even if it meant going into exile. So when the parents eventually came back from abroad, the children were uprooted and dragged off to live with them in a different home. All for their own good, of course.

And indeed it was. Alan enjoyed his holidays in France, quickly picking up French. And his privileged education was eventually to awaken the dormant faculties which he had inherited. His paternal grandfather had been a mathematics scholar at Cambridge, and an Anglo-Irish ancestor on his mother's side had invented the word 'electron' in 1891. (Though it was the fact that he had been elected a Fellow of the Royal Society which impressed the family, making up for the *faux pas* of being a scientist.)

It was all very well young Alan playing at 'stinks' with his schoolboy chemistry set, but this was just a hobby. If he was going to benefit from a proper education, and get on in life, he needed to learn some latin. After all, this was what he was being educated *for*. This was what all the money was being spent on. So he could learn something – not just mess about with his wretched chemistry set. Without latin, he'd never pass the common entrance exam into public school. (In this instance of course, 'common' and 'public' must be understood in precisely the opposite of their usual meaning.)

To the relief of all concerned, Alan managed to scrape into Sherborne, a reasonably prestigious public school in Dorset. At 13 he set off alone from France to begin his first term at his new school. When the ferry arrived at Southampton he discovered the General Strike had begun that very day. The country was at a standstill – there were no trains, or transport of any kind. With typical resourcefulness, he bought a map and set about cycling the 55 miles to Sherborne – where he arrived something of a hero.

But Turing failed to live up to this promise. It soon became obvious that he wasn't the stuff of public school heroes. The scruffy, awkward boy with the stammer didn't seem interested in making friends, or even in making a name for himself. But he wasn't an anti-social rebel either. He simply adopted an asocial approach – preferring as far as possible to go his own way, but *within* the system. This was an attitude which Turing continued to maintain throughout his life. He conformed, but he didn't conform.

In the hearty games-playing ethos of public school, Turing chose the non-team sport of long-distance running. Surprisingly, despite being flat-footed he was good at this. Physically, as well as intellectually, Turing always displayed exceptional endurance – if he was interested. And it was at Sherborne that he discovered a deep interest in mathematics. This developed in typical fashion. He had little interest in pedestrian lessons, and started studying the subject on his own – racing far ahead of the others, whilst at the same time often remaining ignorant of even the most basic principles. He was soon reading about Relativity, and he developed a deep interest in cryptography. He would cut out holes in a sheet of paper: when this was placed over a certain page in a certain book it revealed a message. But creating coded messages requires someone to receive them. At the age of 15, Turing found this someone in an older boy called Christopher Morcom, who was generally reckoned as the best mathematician in the school. Their shared interest in mathematics soon cemented their friendship. For Turing, however, this developed into something more than a friendship. He fell in love with Christopher.

This was an utterly chaste, but utterly bewildering experience for Turing. He couldn't admit his feelings to anyone else, even Christopher (who seemingly remained unaware of the nature of Turing's feelings for him). Turing couldn't even properly admit his feelings to himself. In common with most members of his social class at this time, Turing's ignorance of sex and tender emotions was profound. Smutty jokes and masturbation only exacerbated this ignorance. Sherborne, like other public schools, firmly believed in repression. If sex was ignored, it would go away. Indeed, part of Turing's keenness on cross-country running was that it distracted him from thoughts of masturbation.

Christopher was a fair-headed, blue-eyed boy of slight build. Turing recognised in Christopher a similar inclination to adhere to principles. (These were not particularly evident in Turing, being mainly concerned with the protection of his inner individualism, but they were certainly strong – strong enough to maintain his utter difference from all those around him.) Christopher was a principled character, without being priggish. He didn't join in the 'dirty talk' with the others, which Turing himself found so distasteful.

Christopher was occasionally missing from school, and would return looking wan and thin. Turing knew that Christopher's health was not good, but he had no idea of the seriousness of the situation. Christopher was in fact suffering from bovine tuberculosis. Early in 1931, he was unexpectedly taken ill at school, rushed to hospital in London, and died a few days later.

Turing was devastated. Christopher was the first person to pierce his cocoon of loneliness. He was never to forget his schoolboy love for Christopher. And the idea of chaste unspoken love – often so necessary in those days when you could be jailed for homosexuality – was to see him through many hard times ahead.

Turing duly won a mathematics scholarship to King's College, Cambridge, where he became an undergraduate in October 1931. Initially, he spoke to few people and kept to himself – relishing the novel privacy of a room of his own where he could study in peace. But his stammer spoke volumes, as his psychological turmoil continued. Turing suffered alone and in silence as he came to understand his sexuality – whilst others sneeringly recognised the tell-tale traits of being 'queer'. Fortunately he soon discovered that one of his fellow mathematics scholars shared his inclinations, and they embarked upon a casual sexual affair.

During the early 1930s Cambridge was one of the world's leading mathematical and scientific institutions. The Anglo-Swiss theoretical physicist Paul Dirac and his colleagues had established the university as second only to Göttingen in quantum physics. King's College was particularly favoured – George Hardy, one of the finest mathematicians of his era, and Arthur Eddington, whose work had confirmed Einstein's theory of relativity, were both resident dons and taught Turing.

But Turing's keenest interest was in mathematical logic. In 1913 Russell and Whitehead, both Cambridge dons, had published *Principia Mathematica*. This attempted to provide a philosophical basis for mathematics, establishing its certainty. Russell and Whitehead tried to show that the entire edifice of mathematics could be derived from certain fundamental logical axioms. (In a way, the opposite of Boole's endeavour over half a century previously.) Russell and Whitehead were not entirely successful, as their attempt foundered on certain logical conundrums.

For instance, take such propositions as: 'What I am saying is untrue'. If this proposition is true, what it says is untrue; if it is untrue, what it says is true. In the jargon of the logicians, this proposition was formally indeterminable. Mathematics could not be satisfactorily based on logical axioms until such paradoxes were sorted out. However, many were of the view that such difficulties were superficial. They didn't touch the heart of the project: they didn't invalidate the attempt to base mathematics on a sound logical footing.

All this changed in 1931, when the Austrian *enfant terrible* of logic Kurt Gödel published his paper on the formally indeterminable propositions in *Principia Mathematica.* In this he put forward his notorious proof, which to the horror of his colleague appeared to spell 'the end of mathematics'.

Gödel took the proposition: 'This statement is unprovable', and showed that it cannot be proved true (or this leads to a contradiction), and likewise it cannot be proved false. Gödel managed to demonstrate that within any rigidly logical mathematical system there will always be propositions that cannot be proved, or disproved, according to the axioms upon which that system is based. Mathematics was not complete! Worse still, it appeared to be irreparably flawed. For this meant we couldn't be certain that the basic axioms of arithmetic wouldn't result in contradictions. Mathematics was illogical! (And, horrors of horrors, so too was logic!)

These developments had a profound effect on Turing. Too much, perhaps. For as usual he was way ahead of himself, and neglecting the basics. As a result, he only managed a second class honours in the first part of his Tripos. Fortunately, in his finals he did justice to himself and was awarded a first – which meant he could stay on at Cambridge and do research. Both Eddington and Hardy had no doubts about his exceptional abilities.

By this stage Turing was becoming more confident in himself. He remained a somewhat odd, solitary bird; but he now saw little reason to conceal his homosexuality. Amongst colleagues he would pass the occasional offhand remark about his sexual preferences. This was all part of his principled nature: he insisted upon honesty to himself, and others. There were exceptions however. Mother was not informed about his homosexuality, or his new-found atheism.

Turing got around this deceit with characteristic oddity. At home for Christmas he would sing Easter hymns, and at Easter he would sing

Christmas ones. (Gödel's indeterminacy also had practical applications, it seems.) As Turing's consummate biographer, Andrew Hodges, dryly observes, the family remained 'the last bastion of deceit'.

Meanwhile Alan's mother continued to treat him as the ugly duckling of the family – insisting he smarten up his appearance, ordering him to get a haircut etc, the minute he arrived home in suburban Guildford. (Tax exile abroad had been abandoned in favour of home counties respectability, now that the children's education had been paid for.)

Although Turing was elected a fellow of King's College, and was now one of the most promising mathematical minds in Britain, his mother still felt the need to pass him off rather shamefacedly as a hopelessly head-in-the-clouds child. Turing's boyish appearance only contributed to this. Both physically and mentally he was to retain a curiously youthful demeanour throughout his life.

His relationship with his mother remained close. In his letters home he even tried to keep her informed of his mathematical thinking –

mentioning such things as relativity and quantum theory. How much of this Mrs Turing understood remains open to question. She was an intelligent woman, from an intelligent background – but religiosity and her firm belief in keeping up social appearances remained foremost in her mind. She still regarded Alan as her wayward child. It was typical that he should choose something so *infra dig* as mathematics in which to excel.

But excel he did. His dissertation had secured him a fellowship, and he was now absorbing the latest developments from the finest mathematical and scientific minds of his era. After Hitler took over Germany in 1933, many of the great German exiles passed through Cambridge, often delivering lectures. In this way, Turing was to hear Schrödinger speaking on quantum mechanics, the subject which he had virtually invented. He also heard Max Born, fresh from Göttingen, deliver an entire course on quantum mechanics. Another Göttingen exile Richard Courant gave a course on differential equations.

Both Born and Courant had worked with David Hilbert, the professor of mathematics at

Göttingen, who was generally regarded as one of the greatest mathematicians of all time. Like Russell and Whitehead, he too had tried to establish mathematics on a formal footing, building it up from a few basic axioms. From these, by means of well-defined rules, all mathematical possibilities would follow.

The 'Hilbert programme', as this was known, was also brought to a jarring halt by the so-called 'Gödel catastrophe', which proved that mathematics was logically inconsistent. Yet despite its apparent intent, Gödel's theory did not succeed in destroying mathematics. People went on using it regardless – especially mathematicians. A triangle still seemed to remain a triangle, bridges didn't fall down, and national budgets still seemed to add up (or not: but this was no fault of mathematics). Indeed, many looked upon Gödel's proof as simply an irrelevant interference. Truth was what mattered in mathematics, not consistency. (But are truth and inconsistency compatible?)

Regardless of such wrangling, Gödel's theory still left certain mathematical questions unresolved. And these indicated a way of mitigating the damage. Agreed, an axiomatic system such as mathematics could produce arbitrary propositions (which could neither be proved nor disproved). But was it possible to determine whether such a proposition was arbitrary from within the system? In other words, could such an arbitrary proposition be identified by applying a set of rules derived from the basic axioms upon which the system was based? Could it be determined by a series of given steps – mechanical procedures such as could be followed by anybody, or even a machine? If so, these arbitrary propositions could simply be identified and set aside. They could be ignored, with no overall effect on the system. If however they could not be so identified, all was lost – mathematics would appear to be endemically riddled with inconsistencies.

This was the problem which Turing now set about trying to answer. It was an extremely ambitious undertaking: any solution to it would be fundamental to mathematics. In order to tackle this problem, Turing invented a concept which was to have consequences far beyond mathematics.

What were the mechanical procedures (or rules) which could be used to determine whether or not a mathematical proposition was susceptible to proof? Such rules went to the very heart of calculation. What was a computable number, and how was it calculated? Calculation was a rigid process, such as could be followed by a machine. Turing set about trying to define the theoretical nature of such a machine – now known as a 'Turing machine'.

This machine would operate only according to rules, and was capable of calculating anything for which there was an algorithm – ie, a precise sequence of steps leading to a conclusion.

Take, for instance, the process of discovering a number's factors (ie, the prime numbers by which it is divisible). A simple example: To discover the factors of 180, *divide by the lowest divisible prime number until the prime no longer divides, then repeat this process with the next ascending prime, until the division is complete.* (The prime numbers are those which are only divisible by themselves and one: ie, 2, 3, 5, 7, 11, 13 . . .)

180 ÷ 2 = 90

90 ÷ 2 = 45

$45 \div 3 = 15$

$15 \div 3 = 5$

$5 \div 5 = 1$

Thus $180 = 2^2 \times 3^2 \times 5$

The procedure – or algorithm – is in italics, and can be applied to any number. It can be applied mechanically ie, by a mechanical mind or a mental machine.

Such a machine could be made to follow a certain procedure, and it would then perform a certain task according to the procedural rules. If these rules were for calculating prime numbers, it would calculate prime numbers. If they were the rules of chess, it would be able to play chess. Each machine would simply follow its given procedure.

Turing then posited what he called a 'universal' machine. This could be fed a number which encoded the procedure of any other Turing machine. It would then follow this procedure, and behave in an indistinguishable fashion to the original Turing machine – playing chess, or calculating prime numbers etc.

From this (purely theoretical) standpoint, Turing now set about demonstrating his thesis.

What Gödel had demonstrated was logical. What Turing now proceeded to demonstrate bore a resemblance to Gödel's theory (in its conclusions), but it was *mathematical*.

Turing proposed the concept of a machine capable of recognising arbitrary propositions in a mathematical system. This theoretical machine would need to be a universal Turing machine. It could be fed a number which encoded the description of another Turing machine, and then behave in an identical fashion. But what if this hypothetical universal machine was fed the number which encoded *its own description*. How would it behave as itself, behaving as itself, behaving as itself . . . And how could it follow the procedure to behave as itself when it was already behaving as itself?

Clearly the machine would go mad. In theoretical terms, it was faced with self-contradiction. In other words, such a machine could not exist *even in theory*. Which meant that such a calculation was simply not computable. It was impossible to work out a set of rules which could work out whether a proposition was susceptible to proof (or disproof) using only the procedures from within that system.

Not only was mathematics logically incomplete, as Gödel had shown, it was also mathematically incomplete. There was no mathematical way it could separate itself from its own arbitrary propositions. Turing published his findings in a paper entitled 'On Computable Numbers, with an application to the *Entscheidungsproblem*'. (The latter mouthful refers to the problem of decidability as posed by Hilbert.)

All who even vaguely understood it, recognised that Turing's paper was sensational. (Though few, in that pre-computer era, could possibly have realised that it was epoch-making.) Previously, the central mathematical notions of computability and computable numbers had remained hazy. Now they were clarified. Calculation was defined in precise mathematical terms – so precise that they set out the theoretical blueprint for a machine that could perform the task. At the same time, Turing had defined the limits of *all this machine could do*.

The Turing machine was a theoretical computer. It is now recognised as the theoretical prototype of the electronic digital computer. Turing had mapped out the theory of computers before a single computer (as we know it) had even been constructed.

The task ahead was obvious. But in 1937, when Turing's paper was eventually published, this still lay beyond human capability. (There had even been a delay in publication because no one of suitable calibre could be found to judge the originality of Turing's paper.) By the time 'On Computable Numbers' came out, Turing had crossed the Atlantic and was taking a PhD at Princeton. Here the mathematics department building also housed the recently founded Institute for Advanced Study. (This centre for theoretical scientific research had been established in 1933, and was rapidly becoming the finest of its kind in the world. But in common with several of its great German-Jewish members it remained homeless at this period.) Turing now found himself amongst the gods. Einstein and Gödel were in residence, as well as Courant and Hardy. Most of them remained aloof, barely aware of the young Englishman – who brightly confided to his mother that he was now living the life of a 'confirmed solitary'.

Yet he did make contact with one of the Olympians: the Austro-Hungarian mathematician von Neumann. No scruffy solitary (like Einstein, Gödel, Turing et al), 'Johnny' von Neumann was every inch the Viennese sophisticate, able at the drop of a hat to produce sensational mind-boggling formulae in both mathematics and cocktails alike. But it was von Neumann alone who recognised Turing's full achievement. He understood that the young Englishman had in fact created an entirely new subject. ('Computability' Turing had called it, for want of a word: the subject was so new it still had no name.) It was von Neumann who realised the practical possibilities of this subject. He understood that the next step would be to build a Turing machine.

Meanwhile Turing continued with his PhD, which was related to another of Hilbert's 'problems'. In 1900 Hilbert had outlined 23 important problems for 20th century mathematicians to solve, adding with typical turn-of-the-century positivism 'all mathematical problems are solvable'. Turing had already proved him wrong, but had now decided to make a determined effort to solve a problem related to the Riemann Hypothesis, which Hilbert had called 'the most important in mathematics'. Hardy had already been wrestling with this problem intermittently for 30 years, but to no avail.

Put simply, Turing's related problem concerned the frequency of prime numbers. In the early 1790s the 15-year-old German wunderkind Karl Gauss, seen by some as the only mathematician to match Newton, had discovered that prime numbers appeared to become less frequent according to a regular pattern. For the number *n*, the space between primes increased as the natural logarithm of *n*. This was found to result in only marginal errors. Bernhard Riemann, a successor to Gauss as professor of mathematics at Göttingen, produced an improvement on this which involved the immensely complex Riemann Hypothesis. And even so, Riemann's formula was not precisely correct.

It was found that the logarithm method produced a slight overestimation of the number of primes, and after millions of calculations confirmed this with ever higher numbers, it was assumed that this was always the case. Then one of Hardy's collaborators, J. E. Littlewood, discovered that if the Riemann Hypothesis was true, this was not the case. A cross-over, giving underestimations, took place at some point before the number

$10^{10^{10^{34}}}$

This number is inconceivably large. As Hardy pointed out, it is 'the largest number which has ever served any definite purpose in mathematics'. Even $10^{10^{34}}$ written out in decimal integers would fill books of greater mass than the planet Jupiter, according to Turing's biographer Hodges.

The problems involved here were central to the theory of numbers. Turing struggled manfully through these mental weight-lifting exercises, but to little avail. (The Riemann Hypothesis, for instance, remains unproven to this day.)

Turing found America by turns refreshing and unsettling. He was overworking, spending too much time on his own, and began suffering from bouts of depression. An embarrassing incident involving an ill-judged homosexual advance hardly helped matters. In a letter to his

Cambridge boyfriend (now working as a schoolmaster in Walsall), he mentioned in characteristically offhand fashion that he had worked out a method for committing suicide by eating a lethal apple.

After two years in America, Turing travelled back to Britain, having turned down von Neumann's offer of a place working with him at the Institute for Advanced Study. Turing's fellowship was renewed at King's and he returned to his normal life at Cambridge. He wrote home to Mother for his teddy bear, and avidly attended the opening of *Snow White and the Seven Dwarfs* at the local cinema. He was particularly struck by the scene where the Wicked Witch lowers an apple on a string into the boiling poison, and afterwards took to repeating her chant:

'Dip the apple in the brew
Let the sleeping death seep through.'

Those who heard him chanting this had no idea that he had recently contemplated suicide using an apple. His sexual orientation had accustomed him to living a lie, but he couldn't resist such oblique frankness. (While homosexuality remained illegal, concealment was vital – despite his openness amongst a few Cambridge colleagues.) People continued to find Turing difficult to know; his character was regarded, even then, as an enigma. Yet there *was* a key, if one looked for it. Unfortunately, from now on it would become increasingly difficult to know where to look for this key.

So Turing continued to maintain – indeed insisted upon maintaining – his arbitrary position within the system. He could neither be disapproved (fellow of King's, brilliant mathematician), yet at the same time he could not be approved (by his mother, or on account of his illegal homosexuality).

It was around this time that Turing met the Austrian philosopher Ludwig Wittgenstein, and began attending his lectures. Wittgenstein gave lectures to a select few, who were expected to sit on deck chairs in his bare room and listen while he went about 'thinking' aloud. This often involved long periods of agonising silence, then a question – followed by a savage interrogation of anyone who had the temerity to try and answer it.

Wittgenstein was lecturing on the foundations

of mathematics, but from the philosophical point of view. He was trying to work out the precise nature of mathematics, what exactly it was – rather than how it worked. Turing didn't have Wittgenstein's philosophical brainpower (no one else alive possessed this intimidating force), but he was a better mathematician. And the way Turing saw it, what mathematics was, and what it did, were inextricable. He refused to be cowed by Wittgenstein's brow-beating onslaughts.

At one point Wittgenstein proposed that a system such as mathematics or logic could still remain valid, even if it contained a contradiction. Turing himself had proved that mathematics contained inconsistencies – but these were not quite the same as contradictions. He explained to Wittgenstein that if you tried to build a bridge with mathematics that contained a contradiction, it would fall down. Wittgenstein insisted otherwise: the nature of mathematics and its application were separate matters. But Turing's 'On Computable Numbers' had already shown just how profound was the link between pure and applied mathematics. He had solved a fundamental theoretical problem of mathematics by proposing a 'machine' – which although theoretical, was nonetheless a machine, a practical apparatus which could in principle be built.

Interestingly, Turing's demonstration that any system (such as mathematics or logic) contained undecidable propositions, also disproved Wittgenstein's early philosophy. In this, Wittgenstein had maintained that any problem, as long as it was expressed in proper logical fashion, could always be solved.

Turing was now to enter the realm of applied mathematics with a vengeance. In 1939 war broke out against Nazi Germany, and Turing was seconded to intelligence duties. He was sent to take charge of a code-breaking team in the intelligence establishment at Bletchley Park, 60 miles north of London. This was a top-secret project, under the strictest military surveillance.

The army were unprepared for Turing. Although he still retained his curiously boyish manner, his appearance had by now blossomed into mature Cambridge eccentricity (a condition just this side of clinical oddity).

On first appearance, Turing looked as if he had been sleeping rough. His hair was dishevelled, his nails engrimed, his trousers were held up by an old school tie, and he had given up shaving on a regular basis. (When he cut himself with his razor, a not infrequent occurrence, the sight of blood made him faint.) By this stage Turing's voice had developed into a curious high-pitched upper-class stutter, which was occasionally interrupted by an upsetting nervous laugh (said to have resembled the grating ee-aw of a donkey). When lost in thought – a frequent state – he was liable to accompany his intense mental operations with equally intense squeaks and squawks.

Turing's social attitude was equally upsetting. He simply ignored people whose intellect he didn't rate – which needless to say included the entire military staff who were meant to be running the establishment. And to make matters worse, he tended to work in long bursts lasting for days and nights on end – after which the inspecting officer was liable to come across him playing chess with the office boy, or stretched out for a long afternoon snooze on his desk.

Turing just wasn't amenable to military discipline. Worse still, he didn't appear to take what he was doing seriously. And what he was doing, or *meant* to be doing (as the commanding officer would forcefully remind him), was very serious indeed.

In fact, it was a lot more serious than even the military realised. The efforts of Turing, and the multiplying teams of top-class intelligence workers at Bletchley, almost certainly turned the course of the war.

The story of Bletchley begins in 1938, when a young Polish engineer called Robert Lewinski turned up at the British Embassy in Warsaw. He claimed to have been worked at a factory in Germany, where they were manufacturing code-signalling machines. Lewinski had managed to commit the details of this machine to memory. He was soon smuggled out of Poland to Paris, where he supervised the construction of a machine. The British had heard about these machines, which were known as 'Enigma', and were used by the German command to send coded orders to forces in the field. German U-boat commanders could also use them to identify their position, so they could be sent to the nearest enemy convoys which had been sighted.

The Enigma was astonishingly simple to

operate, yet its encoding system was seemingly all but uncrackable. Basically the system consisted of two machines. The sending machine was set to a key, and the uncoded message was simply typed into it. The message was automatically scrambled by three (or more) electric rotor arms which were set according to the key, and the message was then transmitted. At the other end the receiving Enigma machine was set to the same key, and the message would then be unscrambled and printed out in decoded form. The independently spinning rotors allowed for literally billions of permutations, so that any enemy picking up the scrambled transmission was faced with an apparently impossible task if he wished to decipher the code. Thousands of messages were sent every 24 hours, and the key was changed three times a day. The Germans were justifiably confident that their communications system had an unbreakable code.

Thanks to Lewinski, the British intelligence operators at Bletchley now knew exactly how an Enigma machine was constructed, and how it worked. But this was not enough. Not by a long way. The complexities generated by the Enigma were awesome. Each time a letter was pressed, when typing in a message, the rotors rotated once more. So even if the same letter was struck several times in a row, it would almost invariably produce different letters in the scrambled version. To decipher the code you had to know the key at which the machine had been set: this alone governed the initial position of the rotors. And this could be any one of a million cubed (10^{18}) possibilities when just three rotors were being used. (Top-secret Luftwaffe messages were sent on Enigma machines with *ten* rotor arms.)

Turing and his team (which soon included many of the finest mathematical minds in the country) were faced with a mammoth task. They needed to search through the myriad of coded messages, hunting for any combinations, patterns or possibilities which might somehow mean something, and from this have a stab at working back to the key setting.

Turing assessed the situation at once with typical panache. In theory at least, the problem was quite simple. This was a job for a Turing machine. The machine which Turing had described in his paper 'On Computable Numbers' had not been entirely theoretical. Turing had envisaged a machine which was fed its instructions on a paper tape. The tape was divided into squares, which the machine read one at a time. In its simplest form, any problem could be reduced to a series of instructions in binary digits (bits). As Turing correctly surmised: the problem posed by Enigma was no arbitrary problem. This meant it was solvable – which meant that if the correct instructions were fed into a Turing machine, it could produce the answer. So much for theory: practice was another matter altogether.

Turing and his team set about constructing an electro-magnetic machine which could be instructed to run at high speed through the scrambled Enigma messages, searching for any regularities, recurring features or combinations which might be amenable to decipherment. (Sometimes, as a result of enemy action, they might be able to work out the key to some previous message. This would give a further insight into how Enigma worked, but it would only produce an *out of date* key.) Turing's decoding machine was known as Colossus, and such were the difficulties that no fewer than ten versions of this machine were to be built at Bletchley.

The first Colossus began operating in December 1943. Details of this machine remain sketchy, owing to the obsessive secrecy of the British government and the military. (Until recently, some codes used during the Napoleonic Wars were still listed as classified information.) The Colossus appears to have used 2,400 vacuum tubes computing in binary. It did not contain a stored programme, though it did perform computer-like functions. So was Turing's machine really a Turing machine? This still remains disputed. However, the Colossus is now widely regarded as the forerunner of the electro-magnetic digital computer.

Whatever it was, the Colossus marked a giant leap forward in technology. The combined power of its five processors could scan 25,000 characters a second. But even this was not enough. German U-boats were now sinking ships on the allied Atlantic convoys at an alarming rate. Yet nothing could be done: it was still taking several days to decode the Enigma messages to and from the U-boats. Working day and night, this period was gradually decreased. At one stage Britain only had enough food reserves to last a week. Finally Colossus and the Bletchley team were managing to break the code in hours, then minutes. At last the position of every German U-boat in the Atlantic could be pin-pointed, and the loss of ships in the allied convoys decreased dramatically.

The Germans immediately became suspicious. Despite this, they remained convinced that the Enigma code was unbreakable. The British were obviously receiving information from a network of well-placed spies. There was no need to invent an improved coding machine; the Gestapo were sent in and began making arrests.

Meanwhile Turing continued in his usual fashion, occasionally smartening up for a visit to Mother (who was deeply disappointed that his military duties did not involve a military haircut). Turing's appearance seems to have been indicative of deep psychological uncertainties. He continued his habit of making openly homosexual remarks in front of his colleagues (though this was as far as it went), yet at the same time he also became engaged to one of the female cryptoanalysts (she taught him how to knit gloves). The engagement lasted for six months before he was forced to face up to its futility.

A period of comparative smartness was now followed by another spell of sartorial 'indifference'. Yet despite his slovenly appearance and long exhausting hours of work, Turing remained an exceptionally fit man. Several times a week he would set off on long cross-country runs through the woods and across the fields. Bemused locals would watch as he grasped a passing handful of grass and chewed it as he went along. This was Turing's way of making up for vitamin deficiencies in his wartime diet. (Previously he had always eaten an apple in bed before going to sleep.)

This tendency to idiosyncratic self-sufficiency also extended into unexpected fields. At the outbreak of war Turing had been convinced that Britain would be invaded. He had converted his savings into silver bars, and secretly buried them in the woods near Bletchley Park. He had then encoded their location, and committed it to memory. (Unfortunately, this was the one code which *did* defeat Turing. After the war he couldn't remember it, and never managed to recover his silver bars – despite conducting several systematic and exhaustive 'treasure hunts', and even going so far as to invent his own metal detector.)

The decoding work at Bletchley extended to more than just U-boat locations. Soon almost all German communications were an open book. This work was of such importance that at one stage Turing crossed the Atlantic to liaise with the Americans. During the trip he met von Neumann, who had also begun to put the ideas from 'On Computable Numbers' into practice. At the engineering department of the University of Pennsylvania, the Americans had begun work on ENIAC (the Electronic Numerical Integrator and Calculator). This was even more colossal than Colossus, incorporating an astonishing 19,000 valves. But ENIAC was not to be up and working until after the war. (Unknown to the allies, the Germans were working in this field too. By 1943 Konrad Zuse had produced the first multi-purpose programme-controlled calculator. This was used for analysis in the production of flying bombs, but Zuse's underground laboratory in Berlin was bombed out a year later.)

By the end of the war Turing was working at

Hanslope Park, just down the road from Bletchley, on a speech-coding project named Delilah (after the biblical figure whose deceptive voice had played such havoc on Sampson). By now Turing's work on Colossus had considerably deepened his understanding of electronic machinery. He had begun to ponder on the problem of how machines could be made to imitate the human mind.

In 1945 Turing joined the newly-founded National Physical Laboratory, at Teddington just outside London. Here he headed a project for the construction of an automatic computing engine (known as ACE). Turing set about putting together a design for an electronic digital stored-programme computer. ACE benefitted greatly from Turing's experience in constructing and working with Colossus, but his strong point remained theoretical. Like the universal Turing machine posited in 'On Computable Numbers', ACE was to conform to an overall 'logical design', incorporating many complex logical procedures. Unfortunately these resulted in a number of engineering difficulties, which were of less interest to Turing. His design was way ahead of its time – far superior to ENIAC (the first of the so-called 'von Neumann machines') which was now nearing completion in America, and more advanced than other projects under way in England. But ACE faced more than engineering difficulties. Most notably, it was hampered by a scarcity of funds, and scientific politics.

Unlike other fields of scientific endeavour, scientific politics thrives on lack of funds. Amidst the scarcities of post-war Britain (where even bread was rationed) scientific politics achieved a major historical breakthrough – entering its byzantine era. Such subtle complexities were far beyond the likes of a mere mathematical wizard such as Turing. Never diplomatic at the best of times, Turing soon found his approaches for funds being rebuffed.

The usual reason given for this is that Turing was an uncongenial personality, and his boyish unsartorial appearance meant that people didn't take him seriously. (When attending an outside meeting at some Whitehall department he preferred to run ten miles or so across London, rather than use public transport. Anyone who has witnessed the finish of a cross-country race will be able to visualise the impact of Turing's arrival at a committee of civil servants.) But Turing's lack of PR – for want of a less charitable label – was not the whole story. His was the better scheme, but others schemed better.

By 1947 Turing could see that he was getting nowhere. The official version is that he now resigned from the National Physical Laboratory, leaving others to complete the ACE project. It is unclear whether he jumped, or was pushed.

Ironically, this was the best thing that could have happened to Turing. He went back to Cambridge, where he immediately embarked on revolutionary work in computer theory. Despite his involvement in Colossus, laying the foundations for ACE (which was eventually built, and proved successful), and his future involvement in the development of the computer, it is Turing's theoretical work for which he is best remembered.

As we have seen, Turing had from the start visualised the Turing machine undertaking functions performed by the human mind. But was a machine capable of becoming equivalent to a human mind? Turing now proposed and analysed the concept of 'intelligent machinery'. Moral, human and religious objections were dismissed with characteristic tact: 'Being purely emotional [they] do not really need to be refuted'. Scientific and philosophical objections were more serious. A machine capable of intelligence implied a mechanistic approach to intelligence, which in turn implied determinism. Human intelligence, on the other hand, appeared to include an element of free will.

The tiresome and futile philosophical argument between free will and determinism is irrelevant here. Turing's point was that the human mind *appears* from the outside as if it has free will. It *behaves* as if it does.

So the operations of intelligence are not merely mechanical, but Turing was suggesting that they could be performed by a machine. Surely he was being illogical here? In the verbal sense, perhaps. But Turing had experienced otherwise – during his wartime work on Colossus and Delilah. These were both utterly deterministic machines, yet it had been found that they were also capable of displaying random behaviour. (Not for nothing had Colossus required a team of dozens of 'minders', to keep it following the right track.)

On one level these primitive computer machines had been completely deterministic. Yet on another level they had exhibited distinctly random behaviour which appeared to mimic free will. There was a chink in the armour: a tiny one, but a real one.

The central point made by Turing was that machines could learn. In this way they could extend their operations beyond the merely mechanical. A machine could be taught to improve its behaviour to the point where it exhibited 'intelligence'.

Here Turing overcame another potential objection – which would have limited his thesis. A machine might display intelligence, but this would only be a reflection of the intelligence of its creator. Turing disagreed. He used the analogy of a master and pupil. The pupil could outshine his master, developing a qualitatively superior intelligence, using only the information programmed to him by his master. Turing then argued further. It was possible to produce a machine which played chess (by following the rules which had been fed into it). But 'playing against such a machine gives a definite feeling that one is pitting one's wits against something alive'. Because the computer could learn, its behaviour transcended mechanistic determinism and exhibited an element of freedom which seemed like a living intelligence (which was not necessarily human).

Turing was asking questions which had been asked since the origins of philosophy. What does it mean to be human? What precisely is human intelligence? But he was approaching these questions from an original point of view. Was it possible for a machine to take on these qualities? How do we know an intelligence is human or a machine?

Turing was thinking on a level which transcended mathematics, computable numbers, or even computers. Indeed, he became so involved in his own thought processes that he began to regard the world as if he too was a computer. The mechanics of the computer became the medium of his thought – as he explored the possibilities of thought. What *is* intelligence?

To a certain extent, this identification with a

machine soon began to permeate his whole life. To regard himself as a machine provided a great psychological relief from the continuing turmoil of his inner life.

Turing's return to Cambridge was the best thing that happened to him for both his work and his life. During several periods in Turing's life these two categories remained indistinguishable. This was one where they were not. Turing may have psychologically identified with a computer, but it certainly didn't show in his behaviour.

Turing was now 35, though he still looked around ten years younger. He had rooms in King's College, where his intellect was recognised (by the few who understood) as one of the finest in Cambridge. Alas, his behaviour fell far below this category. Turing took to hanging about his college quadrangle and inviting young men up to his rooms for tea. In the evenings he would call in unexpectedly on other young men in their rooms. As he put it: 'Sometimes you're sitting talking to someone and you know that in three quarters of an hour you will either be having a marvellous night or you will be kicked out of the room.' Identifying with a computer appears to have liberated him from any inhibiting misgivings about his sexuality.

Turing must have been something of a danger to shipping during this period. But fortunately for all concerned, it was not to last. Before his behaviour was deemed by the college authorities to have exceeded the bounds of acceptable eccentricity, Turing found himself a regular boyfriend. This was Neville Johnson, who had won a scholarship from Sunderland Grammar School, and was now in his third year studying mathematics. Neville had already done his two years national service in the army, and Turing was attracted to his rough-and-ready Geordie attitude. Neville Johnson seems to have been one of the very few who penetrated Turing's carapace, his defence against the world. One day, as they lay side by side, Turing confessed: 'I have more contact with this bed, than with other people.' But despite his deep affection for Neville, Turing still occasionally roamed the quad. By this stage complete submission to love was probably impossible. A computer had its intelligence – whether it could possess emotions remained a theoretical question.

Meanwhile big advances were now being made in the practical field. A computing machine known as EDSAC (Electronic Delay Storage Automatic Computer) was in fact being put together at Cambridge, but astonishingly Turing chose to avoid contact with the team responsible. Instead, after a year at Cambridge he took up a post as deputy director of the computer laboratory at Manchester University. Here they were building the Manchester Automatic Digital Machine (popularly known as MADAM).

MADAM was the first electronic stored-programme computer to function. On 21st June 1948, it went into action using its first stored programme – breaking down a number into its constituent prime factors (eg, 4620 = 2^2, 3, 5, 7, 11).

MADAM fulfilled the theoretical specifications of a Turing machine (as described in 'On Computable Numbers'), though it had not been built according to Turing's design. Nonetheless, Turing joined enthusiastically in the expansion of its original faculties. He designed circuits for its input and output hardware, and even obtained from Bletchley the teleprinter of a German encoding machine. Turing was soon putting in long hard hours of mathematical analysis, yet often solving problems with typical flashes of intuitive insight.

Work on MADAM was not all intellect and engineering. Operating this ever-expanding monster of a machine was a Herculean task. According to Turing's assistant: 'Starting in the machine room one alerted the engineer and then used the hand switches to bring down and enter the input programme . . . When this had been achieved one ran upstairs and put the tape in the reader and then returned to the machine room.' If the machine began reading the tape and following the instructions correctly, the operator would call to the engineer for him to switch on the current that activated the writing function. 'As soon as the pattern on the monitor showed that input was ended the engineer switched off the write current . . . it usually took many attempts to get a tape in – each attempt needing another trip to the tape room.' It's lucky Turing remained a fit man.

Despite such athletic difficulties, MADAM was soon able to undertake more complex tasks.

Its tubes were able to store up to 128 40-bit words (groups of binary digits containing instructions which the computer could use). This was not only the first operating computer, but arguably the first to serve a large-scale constructive purpose. Some time later it would be used for calculating the design of the St Lawrence Seaway, one of the great engineering wonders of the 20th century.

However, MADAM's earliest tasks were somewhat less constructive. Turing was more interested in teaching it to play chess, and spent many happy hours at this task, seeking ways to improve its game strategy. Other members of the team were not so happy at the sight of their deputy director and MADAM locked in intellectual combat, and were even less amused by his next venture. Turing programmed MADAM to compose a love letter. This was certainly a first, and produced a missive typical of the awkward passion of one not used to expressing such emotions:

Darling Sweetheart,
You are my avid fellow feeling. My affection

curiously clings to your passionate wish. My liking yearns to your heart. You are my wistful sympathy: my tender liking.
Yours beautifully,
M.U.C.

In this instance MADAM preferred to style herself the Manchester University Computer (MUC) – a choice whose associations might well have interested Sigmund Freud.

Turing's activities here are of equal psychological interest. He had not fallen properly in love since his schoolboy passion with Christopher Morcom. The untimely death of Christopher certainly played a part in his inability to commit himself fully and lastingly to anyone else – though the sheer danger with which homosexual love was fraught during this era in Britain should not be underestimated as a factor here. Either way, Turing was aware of his failing, and suffered from it. (His frank admission to Neville Johnson about how he viewed his past – 'I think of whoever I was in *love* with at the time' – if anything reinforces this view. These love affairs were either thwarted, or culminated in a brief but unsatisfactory conflagration.)

In the light of Turing's computer-identification, his project to programme MADAM to write love letters takes on a particular poignancy. This was no unconscious matter. Turing knew what he was doing, even if the others didn't. By this stage he was blatantly, yet offhandedly open about his homosexual preferences; yet for this very reason his colleagues remained unaware of his secret agony. He may have solved the problem of the Enigma, but he couldn't solve the problem of his own enigma.

Yet even this remained to a large extent a secondary matter – either repressed or ignored. Turing continued to bury himself in his work. (Work and cross-country running remained his bromides.) MADAM's chess-playing and love letters were central to what now most interested Turing. Namely, 'intelligent machinery' – or, as it came to be known: artificial intelligence.

The provocative questions which Turing posed (and often identified with) laid the foundations for this entire field. These questions were profoundly philosophical, without being woolly – and at the same time remained rigidly scientific, without producing mere isolated 'miracles' into which experimental science can so easily degenerate. Like philosophy, but unlike so much of modern science, here was a field of knowledge one could live by: it shed light on the human condition.

Turing put his ideas into a number of papers; the most important of these was 'Computing Machinery and Intelligence', which was published in 1950. In this Turing insisted that computers could be taught how to think for themselves: they were capable of original thought. Widespread opposition to this notion was revealingly dubbed 'sentimental'. So that a computer's processes could be made to resemble the vagaries of human intelligence, Turing suggested the incorporation of a random element like a roulette wheel.

However, many philosophical objections were viewed as tedious and futile. He had no wish for the question of computer intelligence to become bogged down in questions about free will, ethics, the definition of life and so forth. So he brilliantly side-stepped such issues. There was one way to tell whether a machine was intelligent or not:

place it behind a screen and let a human being interrogate it. The human being could then decide, on the basis of written replies, whether he was dealing with an intelligent being or a mere machine. Could a machine fool a human being into thinking it was human? This was the 'imitation game' proposed by Turing (now known as the Turing Test).

Turing showed how a skilful interrogator could probe the machine, eliciting subtle decisions and judgements, and possibly even emotional responses. Or so it would appear from the printed responses. But Turing didn't avoid the philosophical objections altogether (he merely side-stepped the tedious, unproductive ones). His own philosophical argument was all but unanswerable. He insisted that the 'imitation game' must be accepted as a basic criteria. Why? Because this is how we react to *one another*. We have no immediate way of telling whether another person possesses intelligence or not. We can only infer that they are thinking and conscious by comparing them to ourselves. Turing saw no reason why our attitude towards computers should not be the same. 'Why should I be regarded differently from a computer?' he asked. (The fact that this question was asked by a man who did regard himself as a computer opens up a number of intriguing questions. Is there anyone human out there listening to me?)

Magnanimously adopting the human viewpoint, Turing even went so far as to suggest a number of objections to his argument. The most serious of these is now known as 'Lady Lovelace's Objection' – after Babbage's colleague, who first articulated it. Lady Lovelace was convinced that computers were incapable of original thought, because they can only do what they are told to do. In other words, they can only operate within the limits to which we have programmed them.

Turing's reply to this was as calculating as any computer. When we programme a computer, we only have a hazy general idea of what we have set it to do. We certainly haven't worked out all the implications of this task.

By analogy, we have seen that mathematics was once regarded as a simple conception of numbers and a few simple procedures – such as could be fed into a computer. Yet the implications of this system have proved anything but simple. Indeed, they have not only proved all but inexhaustible, they have even evolved their own inconsistency. As Ehrensvard observed: 'There are times when even mathematics seems to have a mind of its own.'

Such thinking eventually led Turing beyond the field of computers into morphogenesis – the development by growth of patterns in organisms. Turing noted that in the same way as mathematics, any simple system grows into complexity. A uniform symmetrical structure develops through diffusion of its form into an asymmetrical structure with a pattern of its own. In 1952 Turing published his initial paper on this subject: 'The Chemical Basis of Morphogenesis.'

This paper poses the question: how do things grow? How does matter *take shape*? (Morphogenesis comes from the ancient Greek words for 'form' and 'origin'.) By co-incidence, Crick and Watson in Cambridge were at that very moment attempting to solve this same problem from a microbiological point of view, eventually coming up with the DNA double helix. But Turing was attacking the problem from a mathematical viewpoint. How did the earth's comparatively simple chemical soup develop into organisms of such immense complexity? Crick and Watson were seeking to discover an explanation of *how* this happened. Turing was seeking an answer for *how* and *why* it happened. He was seeking a mathematical answer which might explain the pattern of life itself in terms of the patterns of mathematics. (If Einstein could explain the ultimate workings of the universe in mathematical formulae, Turing would describe life itself in similar fashion. Turing was nothing if not ambitious.)

How did the primaeval chemical soup contain the information which enabled it to develop complexity? (The parallels with the question of how a computer might develop intelligence are evident.) But what did these problems have to do with mathematics? Various examples pointed to this. Take a saturated inorganic chemical solution in which crystals are forming – or *growing*, which is what they appear to do, developing in an asymmetric and uncannily 'organic' fashion. On the chemical level there is no explanation for any lack of symmetry. But at the molecular level the individual motions and collisions of the molecules in the solution is random. So it is little wonder that the crystals form in an asymmetrical fashion. In a way, the complexity is being *created* at the very moment it occurs.

A revealing example of this process is seen in modern music. The Hungarian composer Georg Ligeti has 'written' a piece for 100 metronomes, all set at different speeds. The metronomes all start at once, and then begin falling out of phase. This sounds like a recipe for chaos – but what in fact develops is a curious 'virtual music', which is in a sense *created by the metronomes themselves*.

Turing was convinced that similar mathematical developments took place in nature. The flowers, plants and cells that he studied all exhibited and developed patterns. Many of these reveal astonishing mathematical sequences.

For instance, the spirals of a fir cone and the packed seeds of a sunflower head both echo the Fibonaci sequence. This is the series 1, 1, 2, 3, 5, 8, 13, 21 . . . where each number is the sum of the two preceding. The intriguing and mysterious properties of Fibonacci numbers echo through mathematics (Pythagorean triangles, prime numbers and the Golden Mean are all related to them) and in nature (pineapples, the growth of leaves, and the distances of the planets from the sun all exhibit Fibonacci features).

The patterns of nature were deeply mathematical. Was it possible that something in the nature of *mathematics* controlled the development of such complexity?

Such were the questions which occupied Turing during the early 1950s. He continued to make use of MADAM during these complex investigations, though he had largely been eased out of further practical development on the computer side. The modern MADAM was expected to do more than write love letters.

By now Turing had bought a house at Wilmslow, a pleasant suburb on the outskirts of Manchester, and was outwardly a figure of some eminence. In 1951 he had been elected a Fellow of the Royal Society at the exceptionally young age of 39. One of his proposers had been the philosopher Bertrand Russell, one of the earliest to recognise the profound *philosophical* import of Turing's work. (This aspect has yet to be fully explored, almost half a century later.)

But respectability was always a very thin

veneer where Turing was concerned. He still worked long hours – he put in regular 12-hour stints at the lab, and it was his habit to 'book' MADAM though the night on Tuesdays and Thursdays. But this left other long, lonely nights when he had no MADAM to occupy him. Instead he would occasionally scout for young homosexual pick-ups.

Something of a friendship developed between Alan and one of his partners, a fair-haired, blue-eyed Manchester lad called Arnold Murray. One weekend Alan left Arnold alone in his house, and on his return he found that it had been burgled. A few minor bits and pieces – including a shirt and a couple of pairs of shoes, some silver fish knives and a compass – were missing. Turing was piqued, and reported the theft to the police.

This proved to be a catastrophic blunder. The investigating detective quickly sniffed out the homosexual element, and in February 1952 Turing was arrested on a charge of 'gross indecency'.

Turing was a resilient character, but the public disgrace inevitably took its toll. He had to travel south to warn his mother of the coming trial and possible publicity. According to his biographer Hodges: 'Mrs Turing was not entirely clear about the significance of what had happened, but she was sufficiently conscious of the subject for there to be a disturbing argument, somewhat at cross purposes.'

Fortunately the trial was not widely reported in the papers. (There was a brief piece in the northern edition of the *News of the World*, under the heading 'Accused Had Powerful Brain'.) The case was almost certainly hushed up by higher authorities. This was perhaps the least they could do for a man who had played a major role in winning the war. One suspects that if Turing had been a more personable chap, and been willing to play the social game, the whole thing might even have been dropped. After all, homosexuality was hardly unknown in the British establishment. But then Turing was resolutely *not* a member of the establishment.

In the end Turing pleaded guilty, and was lucky to escape prison. Instead, he was put on probation – on condition that he underwent a course of hormone treatment to 'cure' him of his condition.

This inept drug treatment had grotesque side-effects. For the time being it rendered him impotent; and as he confided to a pal on a visit to Cambridge: 'I'm growing *breasts*!'

Turing tried burying himself in his work once more. He was now attempting to tackle the questions he had posed in 'The Chemical Basis of Morphogenesis'. But he found himself becoming bogged down in the small variations in first-order systems of differential equations which appeared to give rise to asymmetry. These seemed to account for the chemical theory of morphogenesis when complexity created itself. Complexity creating itself was evidently a complex business.

Turing soon found all this as dispiriting as his failed PhD investigation of prime numbers, in relation to the Riemann Hypothesis. As before, the initial springs of inspiration drained away into a desert of calculations. And once again the possibility of suicide arose.

This time the prospect was more attractive. His work had dried up, and he had by now been excluded from all further creative work with computers – despite his unique qualifications. His sexual identity was virtually obliterated; and the exceptional physique he had maintained as a cross-country athlete had been transformed into a mockery by drugs.

There followed the final enactment of a scene which had certainly been rehearsed on at least one occasion. During the night of 7th June 1954 Alan Turing lay down and ate his customary bedtime apple, which he had treated with cyanide.

AFTERWORD

Turing was largely ignored and forgotten after his death. His wartime work on Colossus remained an official secret, his eventual exclusion from practical creative work on the original British computers meant the victors took the spoils, and his brilliant theoretical work was recognised only by the *cognoscenti*.

All this might have remained the case, but for the brilliant full-scale biography of Turing produced by Andrew Hodges in 1985. This brought Turing the wider public recognition he deserved, as well as exposing an outrageous sexual scandal. (In this case, perpetrated by an ungrateful authority.) All subsequent writers on Turing remain in Hodges' debt. Yet despite exhaustive researches, Turing remained as much an enigma to Hodges as he had been to his contemporaries. Even so, Turing's achievements speak for themselves. He is increasingly recognised as *the* pioneer of computer theory, a founding father of the modern computer – and, almost incidentally, the man who won the war.

The questions of artificial intelligence and morphogenesis – which he was the first to pose in any comprehensive sense – remain central and unanswered to this day.

MAJOR DATES IN THE DEVELOPMENT OF THE COMPUTER

4000BC	Early forms of abacus used in China and Babylon.
1st century BC	A single calculating machine dating from this period has been discovered, and remains a mystery.
1623	Schickard starts making 'calculating clock' in Tübingen: generally recognised as the first digital computer.
1630	Oughtred invents the slide rule: regarded by many as the first analog computer.
1642	Pascal invents a superior calculating machine, capable of handling 8-digit figures.

1673	Leibniz invents a simpler, more efficient calculating machine capable of calculating square roots.
early 19th century	French weaver Jacquard produces cards to control weaving patterns on his looms – the first machine programming.
1823	Babbage starts work on his Difference Engine No. 1.
1854	Boole publishes his paper on binary logic.
1896	Hollerith uses his card-reading machine for US census.
1937	Turing publishes 'On Computable Numbers', outlining the theoretical limits of any future computer.
1948	MADAM becomes the first electronic stored-programme digital computer to go into action.

SUGGESTIONS FOR FURTHER READING

Alan Hodges: *Alan Turing: the enigma* (Vintage) – The brilliant pioneering 580 page biography, still the definitive work on his life.

J. David Bolter: *Turing's Man: Western Culture in the Computer Age* (Duckworth) – Artificial intelligence, how modern computers work, and their social implications.

Joel Shurkin: *Engines of the Mind* (Norton) – The evolution of the computer from mainframes to microprocessors in language which human non-computers can understand.